Standard Review Plan for Decommissioning Cost Estimates for Nuclear Power Reactors

I0494126

Final Report

U.S. Nuclear Regulatory Commission
Office of Nuclear Reactor Regulation
Washington, DC 20555-0001

AVAILABILITY OF REFERENCE MATERIALS
IN NRC PUBLICATIONS

NRC Reference Material

As of November 1999, you may electronically access NUREG-series publications and other NRC records at NRC's Public Electronic Reading Room at http://www.nrc.gov/reading-rm.html.
Publicly released records include, to name a few, NUREG-series publications; *Federal Register* notices; applicant, licensee, and vendor documents and correspondence; NRC correspondence and internal memoranda; bulletins and information notices; inspection and investigative reports; licensee event reports; and Commission papers and their attachments.

NRC publications in the NUREG series, NRC regulations, and *Title 10, Energy*, in the Code of *Federal Regulations* may also be purchased from one of these two sources.
1. The Superintendent of Documents
    U.S. Government Printing Office
    Mail Stop SSOP
    Washington, DC 20402–0001
    Internet: bookstore.gpo.gov
    Telephone: 202-512-1800
    Fax: 202-512-2250
2. The National Technical Information Service
    Springfield, VA 22161–0002
    www.ntis.gov
    1–800–553–6847 or, locally, 703–605–6000

A single copy of each NRC draft report for comment is available free, to the extent of supply, upon written request as follows:
Address: Office of the Chief Information Officer,
          Reproduction and Distribution
            Services Section
          U.S. Nuclear Regulatory Commission
          Washington, DC 20555-0001
E-mail: DISTRIBUTION@nrc.gov
Facsimile: 301–415–2289

Some publications in the NUREG series that are posted at NRC's Web site address http://www.nrc.gov/reading-rm/doc-collections/nuregs are updated periodically and may differ from the last printed version. Although references to material found on a Web site bear the date the material was accessed, the material available on the date cited may subsequently be removed from the site.

Non-NRC Reference Material

Documents available from public and special technical libraries include all open literature items, such as books, journal articles, and transactions, *Federal Register* notices, Federal and State legislation, and congressional reports. Such documents as theses, dissertations, foreign reports and translations, and non-NRC conference proceedings may be purchased from their sponsoring organization.

Copies of industry codes and standards used in a substantive manner in the NRC regulatory process are maintained at—
          The NRC Technical Library
          Two White Flint North
          11545 Rockville Pike
          Rockville, MD 20852–2738

These standards are available in the library for reference use by the public. Codes and standards are usually copyrighted and may be purchased from the originating organization or, if they are American National Standards, from—
          American National Standards Institute
          11 West 42nd Street
          New York, NY 10036–8002
          www.ansi.org
          212–642–4900

NUREG-1713

Standard Review Plan for Decommissioning Cost Estimates for Nuclear Power Reactors

Final Report

Manuscript Completed: October 2004
Published: December 2004

Prepared by:
C.L. Pittiglio

Division of Regulatory Improvement Programs
Office of Nuclear Reactor Regulation
U.S. Nuclear Regulatory Commission
Washington, DC 20555-0001

COMMENTS ON DRAFT REPORT

Standard Review Plans (SRPs) are issued to describe and make available to the public information such as methods acceptable to the NRC staff for implementing specific parts of the NRC's regulations, techniques used by the staff in evaluating specific problems or postulated accidents, and data needed by the NRC staff in its review of applications for permits and licenses. This standard review plan was issued as a draft for public comment in November 2001. Based on use of this document and the public comments provided on the November 2001 version, the SRP has been revised.

This SRP guides the NRC staff in performing a review of each of the decommissioning cost estimates that licensees are required to submit in accordance with 10 CFR 50.75, "Reporting and Recordkeeping for Decommissioning Planning," and 10 CFR 50.82, "Termination of License." The principal purpose of the SRP is to ensure the quality and uniformity of NRC staff reviews and to present a well-defined base from which to evaluate the decommissioning cost estimates that are submitted before decommissioning and at various phases of the decommissioning process. It is also the purpose of the SRP to make the information about regulatory matters widely available so that interested members of the public and the nuclear industry can gain a better understanding of the staff's review process. The SRP identifies the matters to be reviewed, the basis for the review, and the conclusions that are sought.

SRPs are not substitutes for Regulatory Guides or the Commission's regulations, and compliance with them is not required. SRPs are initially issued in draft form for public comment to involve the public in the early stages of developing regulatory positions. Published SRPs will be revised periodically, as appropriate, to accommodate comments and to reflect new information and experience.

ABSTRACT

This Standard Review Plan (SRP) for decommissioning cost estimates provides guidance to Office of Nuclear Reactor Regulation (NRR) and Office of Nuclear Material Safety and Safeguards (NMSS) staff on how to evaluate each of the decommissioning cost estimates that are required to be provided by the power reactor licensees. The SRP includes guidance on evaluating decommissioning costs for both pressurized water reactors (PWRs) and boiling water reactors (BWRs). The SRP is divided into sections that are keyed to the sections in Regulatory Guide-1085, "Standard Format and Content of Decommissioning Cost Estimates for Nuclear Power Reactors," which was developed to provide guidance to licensees on decommissioning cost estimates. Each section of this NUREG is a separate SRP and presents the areas of review, acceptance criteria, review procedures, and evaluation findings for each of the decommissioning cost estimates required by 10 CFR 50.75 and 10 CFR 50.82.

TABLE OF CONTENTS

A.   INTRODUCTION

Decommissioning means permanently removing a nuclear facility from service and reducing radioactive material on the licensed site to levels that permit termination of the NRC license. This Standard Review Plan (SRP) is divided into sections that are keyed to the sections in Regulatory Guide-1085, "Standard Format and Content of Decommissioning Cost Estimates for Nuclear Power Reactors," which is being developed to provide guidance to licensees on decommissioning cost estimates.

NUREG-0586, "Generic Environmental Impact Statement on Decommissioning of Nuclear Facilities, Supplement 1," dated October 2002, evaluated the environmental impact of three methods for decommissioning. The supplemental information to the 1988 decommissioning rule (53 FR 24019) also discussed the three decommissioning methods. A short summary of the three methods follows.

DECON: The equipment, structures, and portions of the facility and site that contain radioactive contaminants are removed or decontaminated to a level that permits termination of the license after cessation of operations.

SAFSTOR: The facility is placed in a safe, stable condition and maintained in that state (safe storage) until it is subsequently decontaminated and dismantled to levels that permit license termination. The determination of SAFSTOR includes those activities necessary for the final decontamination and dismantlement of the facility. During SAFSTOR, a facility is left intact or may be partially dismantled, but the fuel has been removed from the reactor vessel and radioactive liquids have been drained from systems and components and then processed. Radioactive decay occurs during the SAFSTOR period, thus reducing the quantity of contamination and radioactivity that must be disposed of during decontamination and dismantlement (D&D). The definition of SAFSTOR includes the decontamination and dismantlement of the facility at the end of the storage period.

ENTOMB: Radioactive structures, systems, and components are encased in a structurally long-lived substance such as concrete. The entombed structure is appropriately maintained, and monitored until the radioactivity decays to a level that permits termination of the license. Because most power reactors will have radionuclides in concentrations exceeding the limits for unrestricted use even after 100 years and because current regulations require that decommissioning be completed within 60 years of cessation of operation, entombment requests will be handled on a case-by-case basis.

The NRC recognizes that some combination of these methods would also be acceptable. For example, the licensee could conduct a partial radiological decontamination of the plant followed by entombment or a storage period, followed by the completion of the radiological D&D. NUREG/CR-5884 and NUREG/CR-6174 describe two possible scenarios for evaluating the SAFSTOR decommissioning method: SAFSTOR1 and SAFSTOR2. For this SRP, the SAFSTOR2 scenario is assumed where all materials that were originally radioactive still exceed unrestricted release levels and are removed for disposal as low-level waste (LLW). This option results in a more conservative (higher) decommissioning cost estimate than the SAFSTOR1 scenario, which assumes most of the radioactive materials have decayed to unrestricted release levels.

On July 29, 1996, a final rule was published in the *Federal Register* (61 FR 39278) amending the NRC's regulations on the decommissioning procedures that will lead to termination of an operating license for nuclear power reactors. This final rule included changes to 10 CFR Parts 2, 50, and 51.

The revised regulations contain requirements related to decommissioning cost estimates. Regulatory Guide-1085 was written to provide guidance to licensees on the preparation of these cost estimates and to establish a standard format for reporting these cost estimates that is acceptable to the NRC staff.

The guidance in RG-1085 and this SRP apply only to power reactor licensees. The regulations for nonpower reactor licensees are given in 10 CFR 50.82(b).

The minimum decommissioning funding required by the NRC reflects only the efforts necessary to terminate of the Part 50 license. Other activities related to facility deactivation and site closure, including operation of the spent fuel storage pool, construction and operation of an independent spent fuel storage installation (ISFSI), demolition of decontaminated structures, and site restoration activities after residual radioactivity has been removed are not included in the NRC definition of decommissioning. Accordingly, costs for such "nondecommissioning activities" are not addressed in this SRP; however, costs associated with the decontamination of an ISFSI licensed under the general license are included.

B. DISCUSSION

NRC decommissioning funding requirements can be segregated into two categories: (1) those that specify the minimum decommissioning fund that power reactor licensees must obtain and/or maintain to demonstrate reasonable assurance of having adequate funds to decommission their facilities, and (2) those that specify when licensees must submit decommissioning requirements governing site-specific cost estimates. Both sets are relevant to this SRP and are discussed below.

1.   FINANCIAL ASSURANCE

Licensees of operating nuclear power reactors must provide reasonable assurance that funds will be available for the decommissioning process. For these licensees, reasonable assurance consists of fulfilling a series of steps identified in 10 CFR 50.75(b), (c), (e), and (f). These steps assure that the licensee can certify that financial assurance is in effect for an amount that may be more but not less than the amount stated in the table in 10 CFR 50.75(c)(1). Specifically, this table states that if P equals the thermal power of a reactor in megawatts (MWt), the minimum financial assurance (MFA) funding amount in millions of January 1986 dollars is:

(1) For a PWR: $MFA = (75 + 0.0088P)$

(2) For a BWR: $MFA = (104 + 0.009P)$

For either a PWR or BWR, if the thermal power of the reactor is less than 1200 MWt, then the value of P to be used in 1 and 2 is 1200, and if the thermal power is greater than 3400 MWt, then a value of 3400 is used for P. That is, P is never less than 1200 nor greater than 3400. The financial assurance amounts calculated in equations 1 and 2 are based on January 1986 dollars, in millions. To account for inflation from 1986 to the current year, these amounts must be adjusted annually by multiplying 1 and 2 by an escalation factor (ESC) described in 10 CFR 50.75(c)(2). This ESC is

$$ESC \text{ (current year)} = (0.65L + 0.13E + 0.22B)$$

where L and E are the ESCs from 1986 to the current year for labor and energy, respectively, and are to be taken from regional data of U.S. Department of Labor, Bureau of Labor Statistics, and B is an annual ESC from 1986 to the current year for waste burial and is to be taken from the most recent revision of NUREG-1307, "Report on Waste Disposal Charges: Changes in Decommissioning Waste Disposal Costs at Low-Level Waste Burial Facilities." NUREG-1307 is updated from time to time to account for disposal charge changes. In January 1986 (the base year), using disposal costs from DOE's Hanford Reservation waste disposal site, L, E, and B all equaled unity; thus the ESC itself equaled unity. A discussion of the origin of the $0.65L$, $0.13E$, and $0.22B$ terms is given in NUREG-1307. Thus,

$$MFA \text{ (in millions, current year dollars)} = MFA \text{ (in millions, 1986 dollars)} \times ESC \text{ (current year)}$$

NUREG-1307 provides several examples of how to determine the minimum decommissioning fund requirement using the above algorithm.

2. DECOMMISSIONING COST ESTIMATES

The regulations summarized below apply to decommissioning cost estimates:

- 10 CFR 50.75(f)(2) requires that a licensee "...shall at or about 5 years prior to the projected end of operations submit a preliminary decommissioning cost estimate (herein after referred to as the preliminary cost estimate) which includes an up-to-date assessment of the major factors that could affect the cost to decommission." Section 50.75(f)(4) requires a licensee to include plans to adjust funding levels to demonstrate a reasonable level of financial assurance, if necessary, in the preliminary cost estimate.

  In addition, 10 CFR 50.75(c) specifies that the initial certification amount of funds for decommissioning be based on the amounts specified in 10 CFR 50.75(c), which represent the minimum funding level that applicants and licensees must meet. However to meet the 10 CFR 50.75(c) requirements, a power reactor licensee may submit a certification based on a site-specific cost estimate which may be more but not less than the 10 CFR 50.75(b)(1) estimate when a higher funding level is desired than that provided in 10 CFR 50.75(c). The basis for any increases should be provided.

- 10 CFR 50.82(a)(4)(i) requires a licensee to provide an estimate of expected costs for the activities being proposed in the Post-Shutdown Decommissioning Activities Report (PSDAR). The PSDAR is to be submitted prior to or within 2 years following permanent cessation of operations. Regulatory Guide 1.185, "Standard Format and Content for Post-Shutdown Decommissioning Activities Report," identifies the type of information in the PSDAR that would be acceptable to the NRC staff. The cost estimate may be the amount of decommissioning funds estimated to be required

3

pursuant to 10 CFR 50.75(b) and (c) as currently reported on a calendar-year basis at least once every 2 years to the NRC according to 10 CFR 50.75(f)(1), or a site-specific cost estimate.

- 10 CFR 50.82(a)(8)(iii) requires a licensee to provide a site-specific decommissioning cost estimate within 2 years following permanent cessation of operations. This requirement may be satisfied by including a site-specific estimate as part of the PSDAR. In addition, 10 CFR 50.75(c) specifies that the initial certification amount of funds for decommissioning be based on 10 CFR 50.75(c)(1), which represent the minimum funding level that licensees must meet. The site-specific cost estimate submitted within 2 years following permanent cessation of operations may be significantly higher than the funding level based on the formula. If the site-specific cost estimate results in a funding level that differs from the amount specified in 10 CFR 50.75(c), the licensee must provide the basis for the change.

- 10 CFR 50.82(a)(9)(ii)(F) requires that a licensee provide "an updated site-specific estimate of remaining decommissioning costs..." as part of a License Termination Plan (LTP). According to 10 CFR 50.82(a)(9)(i), the licensee must submit the LTP at least 2 years before termination of the license.

As provided in 10 CFR 50.82(a)(8)(ii), a licensee may at any time without prior notification to the NRC withdraw funds from the decommissioning trust up to a cumulative total of 3 percent of the generic amount calculated under 10 CFR 50.75 for decommissioning planning purposes. After submittal of the certifications of permanent shutdown and fuel removal required under 10 CFR 50.82(a)(1) and commencing 90 days after the NRC has received the PSDAR, the licensee may use an additional 20 percent of the decommissioning funds prescribed in 10 CFR 50.75(c) for decommissioning purposes. The licensee is prohibited from using the remaining 77 percent of the generic decommissioning funds until a site-specific decommissioning cost estimate (SSCE) is submitted to the NRC. In addition, use of decommissioning funds is limited by 10 CFR 50.82(a)(8)(i) to legitimate decommissioning expenses that neither reduce the value of the trust fund below the amount necessary to place and maintain the reactor in a safe storage condition, nor inhibit the licensee's ability to completely fund the trust so that the site is released the license . terminated.

3.   DECOMMISSIONING COST DEFINITION

As defined in 10 CFR 50.2, "*Decommission* means to remove a facility or site safely from service and reduce residual radioactivity to a level that permits—

(1) Release of the property for unrestricted use and termination of the [Part 50] license; or

(2) Release of the property under restricted conditions and termination of the [Part 50] license."

The decommissioning cost estimates required by the regulations referenced above apply only to those costs that necessary to accomplish the purposes listed in the definition above. Costs that may be incurred by a licensee when it removes a facility from service or restores the site after decontamination is complete but that do not reduce residual radioactivity or are not required to terminate the license are not considered NRC decommissioning costs. Accordingly, they should not be included in the NRC decommissioning cost estimate. A licensee may choose to report non-NRC decommissioning costs along with its decommissioning cost estimate; however, such costs need to be clearly identified and separated.

4. COST ADJUSTMENT METHODOLOGY

The decommissioning cost estimates based on 10 CFR 50.75(c) for the reference PWR and reference BWR presented in this SRP are based on information developed in NUREG/CR-5884 and NUREG/CR-6174, respectively. All costs presented in this SRP include a 25% contingency factor and are in year 2000 dollars. The cost adjustment methodology described in this section can be used to adjust the costs in this report from year 2000 dollars to any future year. As discussed in Section B.1, costs are divided into three general areas that tend to escalate similarly: (1) labor, materials, and services, (2) energy and waste transportation, and (3) radioactive waste burial/disposition. A typical allocation of cost adjustment factors to the set of reference reactor cost components is presented below in Table 1.

A relatively simple equation can be used to estimate decommissioning costs to account for escalation from the base year 2000 to any other year of interest, year(x). That equation is

$$Estimated\ cost\ [year(x)] = A_{base}\ L_x + B_{base}\ E_x + C_{base}\ B_x$$

A_{base} = sum of all labor, material, and services cost components

L_x = labor, material, and services adjustment factor, base year 2000 to year(x)

B_{base} = sum of all energy and transportation cost components

E_x = energy and transportation adjustment factor, base year 2000 to year(x)

C_{base} = sum of all radioactive waste burial/disposition costs components, and

B_x = radioactive waste burial/disposition adjustment factor, base year 2000 to year(x)

Table 1. Cost Adjustment Factors Used for Decommissioning Cost Estimates of the Reference PWR [a] and Reference BWR [b]

PWR Cost Component	Adjustment Factor Used	BWR Cost Component	Adjustment Factor Used
Radioactive Component		**Radioactive Component**	
Removal of RPV Internals	L_x	RPV Internals	L_x
Removal of Reactor	L_x	Reactor Pressure Vessel	L_x
Steam Generator Removal	L_x	Sacrificial Shield	L_x
Generator Clading Costs	L_x	Recirculation Pumps	L_x
RCS Piping	L_x	RCS Piping	L_x
Large Miscellaneous RCS	L_x	RCS Piping Insulation	L_x
Small Miscellaneous RCS	L_x	Main Turbine	L_x
Pressurizer	L_x	Main Turbine Condenser	L_x
Pressurizer Relief Tank	L_x	Moisture Separator	L_x
Primary Pumps	L_x	Feed Water Heaters	L_x
Spent Fuel Racks	L_x	Turbine Feed Pumps	L_x
Biological Shield	L_x	Structural Beams, Plates, &	L_x
		Spent Fuel Racks	L_x
Decon. & Dismantlement			
Decon. Buildings	L_x	**Decon. & Dismantlement**	
Removal of Plant Systems	L_x	Decon. of Buildings	L_x
		Removal of Plant Systems	L_x
Management and Support			
Support Staff	L_x	**Management and Support**	
DOC Staff	L_x	Support Staff	L_x
Consultant/Other Staff	L_x	DOC Staff	L_x
Termination Survey Costs	L_x	Consultant/Other Staff	L_x
Regulatory Costs	L_x	Termination Survey Costs	L_x
Special Tools & Equipment	L_x	Regulatory Costs	L_x
Monitoring Costs	L_x	Special Tools and	L_x
Laundry Services	L_x	Environmental Monitoring	L_x
Maintenance Allowance	L_x	Laundry Services	L_x
Small Tools & Equipment	L_x	Maintenance Allowance	L_x
Nuclear Liability Insurance	L_x	Small Tools or Equipment	L_x
Property Taxes	L_x	Nuclear Liability Insurance	L_x
DOC	L_x	Property Taxes	L_x
Steam	L_x	DOC	L_x
Chemical Decon	E_x	Chemical Decontamination	E_x
Plant Power Usage	E_x	Plant Power Usage	E_x
LLW Packaging	L_x	**LLW Packaging**	L_x
LLW Shipping	E_x	**LLW Shipping**	E_x
LLW Burial/Waste Vendor	B_x	**LLW Burial/Waste Vendor**	B_x

[a] NUREG/CR-5884
[b] NUREG/CR-6174

4.1   Labor Adjustment Factors

The adjustment factor for labor, L_x, can be obtained from the "Monthly Labor Review," published by the U.S. Department of Labor, Bureau of Labor Statistics (BLS). Specifically, the appropriate regional data from the table (currently Table 24) entitled "Employment Cost Index, Private Nonfarm Workers, by Bargaining Status, Region, and Area Size," subtitled "Compensation," should be used. These labor adjustment factors can also be obtained from BLS databases made available on the World Wide Web (see NUREG-1307, Appendix C, for instructions). L_x should be adjusted from a base value in Table 24 corresponding to base year 2000, to the year(x) of interest.

To calculate a labor adjustment factor for a particular region, two indices are needed, a value for the base year and a value for the year (x) of interest. These values are shown in Table 2 for each region. The base year 2000 values of L_x from the BLS data are provided in column 2 of Table 2. To adjust the costs to a future year(x), the year (x) values for L_x from the BLS data should be substituted in column 3 (year (x) of interest).

Table 2. Labor Cost Adjustment Factors by Region

Region	Base Year (2000)	Year (x) of Interest
Northeast	144.3	$x_{Northeast}$
South	143.0	x_{South}
Midwest	146.3	$x_{Midwest}$
West	144.7	x_{West}

In general, L_x is calculated for each region by dividing the Year (x) of Interest value (column 3) by the Base Year 2000 value (column 2).

Future labor adjustment factors from BLS should be treated similarly. Future revisions to NUREG-1307 will provide new base year calculations as appropriate. However, if BLS has changed its base year and the change is not reflected in the current revision of NUREG-1307, the licensee should calculate the labor adjustment factor to reflect applicable changes.

4.2   Energy Adjustment Factors

The adjustment factor for energy, E_x, can be obtained from the "Producer Price Indexes," published by the U.S. Department of Labor, Bureau of Labor Statistics (BLS). Specifically, data from the table (currently Table 6) entitled "Producer Price Indexes and Percent Changes for Commodity Groupings and Individual Items" (PPI) should be used.

E_x consists of two components, industrial electric power, P_x, and light fuel oil, F_x. Hence, E_x should be obtained using the BLS data in the following equations:
   for the reference PWR: $E_x = [0.58P_x + 0.42F_x]$

7

for the reference BWR: $E_x = [0.54P_x + 0.46F_x]$

These equations are derived from Table 6.3 of NUREG/CR-0130 and Table 5.3 of NUREG/CR-0672. P_x should be taken from data for industrial electric power (Commodity code 0543), and F_x should be taken from data for light fuel oils (Commodity code 0573). These energy adjustment factors can also be obtained from BLS databases made available on the World Wide Web (see NUREG-1307, Appendix C, for instructions). The Base Year 2000 values for P_x and F_x from BLS data are provided in column 2 of Table 3.

Table 3. Energy Cost Adjustment Factors by Energy Source

	Base Year (2000)	Year (x) of Interest
Industrial electric power	126.5	$X_{electric}$
Light fuel oils	72.9	$X_{fuel\ oil}$

As discussed for L_x in Section 3.1 above, to adjust the costs to a future current year (x) , the year (x) values for P_x and F_x should be substituted in column 3. The base year 2000 values of P_x and F_x from the BLS data are 126.5 and 72.9, respectively. No regional BLS data for these PPI commodity codes are currently available. Thus, the values of P_x and F_x for the year (x) of interest are:

$$P_x = (X_{electric})_{Year(x)\ of\ interest} \div (126.5)_{Base\ Year\ 2000}$$

$$F_x = (X_{fuel\ oil})_{Year(x)\ of\ interest} \div (72.9)_{Base\ Year\ 2000}$$

The value of E_x for the reference PWR is therefore

$$E_x = [(0.58P_x) + (0.42F_x)]$$

This value of E_x should then be used in the equation to adjust the energy costs to year(x) dollars for decommissioning a PWR. Correspondingly, the value of E_x for the reference BWR is:

$$E_x = [(0.54P_x) + (0.46F_x)]$$

Future energy adjustment factors from BLS should be treated similarly. Future revisions to NUREG-1307 will provide new base year calculations as appropriate. However, if BLS has changed its base year, and the change is not reflected in the current revision of NUREG-1307, the licensee should calculate the energy adjustment factor to reflect applicable changes.

4.3 Waste Burial Adjustment Factors

The adjustment factor for waste burial/disposition, B_x, can be taken directly from data for the appropriate LLW burial location as given in Table 2.1 of the most recent revision of

NUREG-1307. For example, $B_x = 18.129$ (in 2000 dollars) for a PWR directly disposing all decommissioning LLW at the South Carolina burial site. The base year 2000 values for B_x are provided in columns 2 and 3 of Table 4.

Table 4. Waste Burial/Disposition Cost Adjustment Factors by Disposition Option and Site

Waste Burial	Base Year (2000)		Year(x) of Interest	
	PWR	BWR	PWR	BWR
Direct Disposal/WA [a]	2.223	3.375	$X_{PWR\ Direct\ Disposal/WA}$	$X_{BWR\ Direct\ Disposal/WA}$
Direct Disposal/SC [b]	18.129	16.244	$X_{PWR\ Direct\ Disposal/SC}$	$X_{BWR\ Direct\ Disposal/SC}$
Waste Vendor/WA	4.060	4.379	$X_{PWR\ Waste\ Vendor/WA}$	$X_{BWR\ Waste\ Vendor/WA}$
Waste Vendor/SC	8.052	8.189	$X_{PWR\ Waste\ Vendor/SC}$	$X_{BWR\ Waste\ Vendor/SC}$

[a] WA refers to the Washington LLW disposal site located near Richland, Washington.
[b] SC refers to the South Carolina LLW disposal site located near Barnwell, South Carolina.

As discussed for L_x and E_x above, to adjust the costs to a future Year (x), the Year (x) values for B_x from the latest revision of NUREG-1307 should be substituted in columns 4 and 5 of Table 4. For example, to adjust waste disposal costs using the waste vendor option for LLW from a PWR at the South Carolina disposal site from base year 2000 (basis for this SRP) to the waste vendor option at the Washington disposal site in Year (x):

$$B_x = (X_{PWR\ Waste\ Vendor/WA})_{year(x)\ of\ interest} \div (8.052)_{base\ year\ 2000}$$

This value of B_x should then be used in the equation to adjust the waste burial cost to year (x) dollars for LLW waste disposition from a PWR using the waste vendor option with the Washington disposal site.

C. STANDARD REVIEW PLAN FOR DECOMMISSIONING COST ESTIMATES

The purpose of this SRP is to direct the NRC staff's review of the licensee's cost estimates. The major types of cost estimates affecting the licensee are the preliminary cost estimate, the estimate of expected costs presented in the PSDAR, the SSCE required within 2 years following permanent cessation of operations, and the updated SSCE required as part of the LTP. In addition, a licensee may submit a certification amount of funds for decommissioning based on an SSCE that is equal to or greater than that calculated in the formula in 10 CFR 50.75(c)(1) or (2) when a higher funding level is desired. Individual SRPs are provided for the preliminary cost estimate, the estimate of expected costs presented in the PSDAR, the SSCE, and the updated SSCE.

Each SRP is divided into the following sections: (1) Review Responsibilities, (2) Areas of Review, (3) Acceptance Criteria, (4) Review Procedures, (5) Evaluation Findings, and (6) Implementation.

1. PRELIMINARY COST ESTIMATE

The preliminary cost estimate is required at or about 5 years prior to the projected end of operations. The projected end of operations need not be the same as the expiration date of the operating license if a licensee chooses to permanently cease operations at an earlier date. In some cases, a licensee may prematurely shut down and submit its certification of permanent cessation of operations, as required by 10 CFR 50.82(a)(1), more than 5 years prior to the expiration date of the operating license. In this event, the requirement of 10 CFR 50.75(f)(2) to submit a preliminary cost estimate is not applicable. A licensee could choose to submit its preliminary cost estimate as the estimate of expected costs presented in the PSDAR, and thereby satisfy the requirements of 10 CFR 50.82(a)(4)(i).

According to 10 CFR 50.75(f)(4), the licensee is required to include in the preliminary cost estimate plans for adjusting levels of funds for decommissioning, if necessary to demonstrate a reasonable level of assurance that funds will be available when needed to cover the costs of decommissioning. The reviewer should determine whether the licensee must comply with this requirement. If it is required, the reviewer should determine whether the plans provide adequate financial assurance.

By 10 CFR 50.82(a)(8)(iv), licensees who plan to use a period of storage or surveillance (SAFSTOR) are required to provide a means of adjusting cost estimates and associated funding levels over the period of storage or surveillance. If a licensee plans to use a period of SAFSTOR, the reviewer should ensure that the licensee has included a description of its means of adjustment with its preliminary cost estimate. The reviewer should determine if the means described by the licensee provides adequate assurance that funds will be available for decommissioning activities at the time they are needed.

1.1 Review Responsibilities

Primary— Cognizant Project Manager, Project Directorate, Division of Licensing Project Management, Office of Nuclear Reactor Regulation, or as assigned

Secondary— Financial Reviewer, Financial and Regulatory Analysis Section, Reactor Policy and Rulemaking Branch, Division of Regulatory Improvement Programs, Office of Nuclear Reactor Regulation, or as assigned

1.2 Areas of Review

This SRP directs the staff's review of the preliminary cost estimate that 10 CFR 50.75(f)(2) requires to be submitted at or about 5 years before the projected end of operations. The intent of this preliminary estimate is to provide the NRC with an up-to-date estimate of expected costs and identify major factors in the cost of the decommissioning. The licensee will have already submitted a cost estimate for establishing a fund for decommissioning as required by 10 CFR 50.75(b). This estimate will have been revised periodically during operation and may be used in preparing the preliminary cost estimate. The preliminary cost estimate will generally be substantially less detailed than the SSCE.

The scope of the review directed by this SRP includes (1) a comparison of the preliminary cost estimate with the minimum decommissioning funding required, and (2) an assessment of the major factors that could affect the preliminary cost estimate.

1.3   Acceptance Criteria

The acceptance criteria are based on the requirements of 10 CFR 50.75(f)(2), 10 CFR 50.75(f)(4), and 10 CFR 50.82(a)(8)(iv), as applicable. The regulations require that each power reactor licensee shall at or about 5 years prior to the projected end of operations submit a preliminary cost estimate which includes an up-to-date assessment of the major factors that could affect the cost to decommission.

- The reviewer should compare the preliminary cost estimate to the minimum decommissioning funding required under 10 CFR 50.75(b) to ensure that the licensee's submittal meets the intent of the regulations given in 10 CFR 50.75.

- The reviewer should ensure that the preliminary cost estimate includes an up-to-date listing of the major factors that could affect the cost to decommission and that these factors are assessed by the licensee.

1.4   Review Procedures

The reviewer will use the following process to determine that the cost estimate has been submitted and that the estimate included an up-to-date assessment of the major factors that could affect the cost to decommission.

1.4.1   Comparison of the preliminary cost estimate to the minimum required decommissioning fund

The reviewer should calculate the minimum decommissioning financial assurance requirement amount derived per the algorithm discussed in Section B.1 of this SRP (10 CFR 50.75(c)) and compare it to the preliminary cost estimate amount. The preliminary cost estimate is acceptable if it is greater than or equal to the decommissioning financial assurance requirement amount. If the preliminary cost estimate is less than the amount derived from the algorithm in 10 CFR 50.75(c), the reviewer shall provide this information to the NRC project manager who will document the finding and inform the licensee in writing of additional information needed to resolve the deficiency.

If the preliminary cost estimate differs from the amount of the generic decommissioning fund amount of 10 CFR 50.75(c), the reviewer should assess the licensee's cost estimate to determine whether all significant costs have been included. The reviewer should assess site-specific conditions identified by the licensee to determine if the site-specific conditions would significantly impact the amount calculated in accordance with 10 CFR 50.75(c).

1.4.2   Assessment of the major factors that could affect the preliminary cost estimate

The following factors should be used by the reviewer to ensure that the cost estimate includes an up-to-date assessment of the major factors that could affect the cost to decommission:

- the decommissioning option/method anticipated to be used
- the potential for known or suspected contamination of the facility or site to affect the cost of decommissioning
- the LLW disposition plan
- the preliminary schedule of decommissioning activities
- any other factors that could significantly affect the cost to decommission

The reviewer should review the preliminary cost estimate to determine if it is sufficiently detailed to allow the reviewer to assess its adequacy. To make this assessment, the reviewer should confirm that the cost estimate is provided in current year (estimate year) dollars and that it accounts for the entire decommissioning work scope. The cost estimate should provide costs for each of the following, or similar, major decommissioning phases:

- Pre-decommissioning engineering and planning—decommissioning engineering and planning prior to completion of reactor defueling

- Reactor deactivation—deactivation and radiological decontamination of plant systems to place the reactor into a safe, permanent shutdown condition

- Safe storage—safe storage monitoring of the facility until dismantlement begins (if storage or monitoring of spent fuel is included in the cost estimate, it should be shown separately)

- Dismantlement—radiological decontamination and dismantlement (D&D) of systems and structures required for license termination (if demolition of uncontaminated structures and site restoration activities are included in the cost estimate, they should be shown separately)

- Low-level radioactive waste (LLW) disposition—LLW packaging, transportation, vendor processing, and disposal. Tables 5 and 6 provide decommissioning cost estimates by these major activities for the NRC reference PWR[1] (NUREG/CR-5884) and reference BWR[2] (NUREG/CR-6174), respectively. The reviewer should compare

---

[1] The Portland General Electric Company's Trojan nuclear plant, at Rainier, Oregon, is used as the reference PWR power station. Trojan is an 1175-MW(e) single-reactor power station that utilizes a four-loop pressurized water reactor manufactured by the Westinghouse Electric Corporation in the nuclear steam supply system. Although Trojan was prematurely shutdown on January 4, 1993, the reevaluated decommissioning cost analyses assumed that the Trojan plant operated for the full term of its license to be more representative of large PWRs in general.

[2] The Washington Public Power Supply System's Washington Nuclear Plant Two (WNP-2) at Richland, Washington, is used as the reference BWR power station. WNP-2 is an 1155 MW(e) single-reactor power station that utilizes a nuclear steam supply system with a direct-cycle boiling water reactor

the pliminarycost estimate with the cost values provided in Tables 5 and 6 to make a judgment of the reasonableness of the preliminary cost estimate, recognizing the differences between the reactor for which the preliminary cost estimate was developed and the reference reactors.

If necessary, as required by 10 CFR 50.75(f)(4), the preliminary cost estimate shall also include plans for adjusting levels of funds assured for decommissioning to demonstrate a reasonable level of assurance that funds will be available when needed to cover the cost of decommissioning. However, the evaluation of the reasonable assurance of funding is not conducted as part of the review of the licensee's decommissioning cost estimate. It is conducted according to NUREG-1577. The reviewer should ensure that the appropriate information has been provided.

The reviewer should confirm that the licensee has taken into account any major factors that could affect the cost to decommission. Major factors include the following:

- The decommissioning option/method anticipated to be used. The decommissioning options generally available are DECON, SAFSTOR, or some combination thereof. Section A of this SRP describes each of these options. If the chosen option/method will result in completion of decommissioning more than 60 years after cessation of operations, identification and assessment of the factors causing this delay should be presented. Acceptable factors from 10 CFR 50.82(a)(3) include unavailability of waste disposal capacity and other site-specific factors, such as the presence of other nuclear facilities at the site.

- The potential for known or suspected contamination at the site. Although the requirements described in 10 CFR 50.75(g) for keeping records of spills or other unusual occurrences are outside the scope of this SRP, the reviewer should ensure that the licensee has evaluated the anticipated extent of contamination on the facility and site based on information available in the decommissioning files. This description need not be a detailed discussion but should include descriptions of known instances of releases of contaminated materials into the facility and the external environment, and the possible impact on decommissioning. Known environmental contamination should be identified (including soil, groundwater, surface water, etc.). (Note, the files required to be kept, pursuant to 10 CFR 50.75(g), include records of spills or other unusual occurrences involving the spread of contamination in and around the facility, equipment, or site; as-built drawings and modifications of structures and equipment in restricted areas where radioactive materials are used and/or stored and of locations of possible inaccessible contamination such as buried pipes which may be subject to contamination; records of the cost estimates performed for the decommissioning funding plan or of the amount certified for decommissioning; and records of the funding method used for assuring funds if either a funding plan or certification is used.)

- A brief description of the plans for LLW disposal. The reviewer should determine if the licensee specifically evaluated the plans for LLW management, including the anticipated LLW disposal situation, and how LLW will be managed if no LLW disposal

---

manufactured by the General Electric Company. WNP-2 has a Mark II containment. The reevaluated decommissioning cost analyses assumed that the WNP-2 plant operated for the full term of its license.

sites are available. The reviewer should understand the site-specific factors that could impact the disposition of spent fuel and LLW to determine the reasonableness of these plans.

- A preliminary schedule that shows the major decommissioning activities and the time period over which each of these activities extend. Typical major decommissioning activities were described above.

- Any other major site-specific factors that could have a significant effect on the cost of decommissioning, such as large volumes of mixed radioactive-hazardous wastes with uncertain disposition pathways and known regulatory or technical issues having uncertain resolution outcomes.

1.5 Evaluation Findings

Using the acceptance criteria in C.1(3) and the review procedure in C.1(4) of this section as a basis, the NRC reviewer shall verify that sufficient information has been provided to satisfy the requirements of the underlying regulations (10 CFR 50.75(f)(2)). The preliminary cost estimate shall be considered deficient if the decommissioning cost estimate is less than the financial assurance amount required by 10 CFR 50.75(c), or if the assessment of the major factors that could affect the preliminary cost estimate are not adequate, or if site-specific factors invalidate the technical basis of the formula used to calculate the minimum fund amount in 10 CFR 50.75(c). If deficiencies are discovered, the reviewer should request the appropriate information from the licensee in writing. The reviewer documents the findings of his/her review of the preliminary cost estimate and places a copy of the memorandum into the licensee's docket.

If the licensee included plans to adjust the level of funds assured for decommissioning in accordance with 10 CFR 50.75(f)(4) and 10 CFR 50.82(a)(8)(iv), the reviewer should document the plans to adjust the level of funding.

1.6 Implementation
The method described in this SRP will be used by the staff in evaluating conformance with the Commission's regulations, except when the licensee proposes an acceptable alternative for complying with specified portions of the regulations.

2. ESTIMATE OF EXPECTED COSTS IN THE PSDAR

Prior to or within 2 years following permanent cessation of operations, the licensee is required by 10 CFR 50.82(a)(4)(i) to submit a PSDAR to the NRC. In addition to other prescribed content, this report is required to include an estimate of expected costs. Regulatory Guide 1.185 identifies the type of information to be contained in the PSDAR that would be acceptable to the NRC staff. The cost estimate may be the amount of decommissioning funds estimated to be required by 10 CFR 50.75(b) and (c) as currently reported on a calendar-year basis at least once every 2 years to the NRC according to 10 CFR 50.75(f)(1), or it may be a site-specific cost estimate. Other related but non-NRC decommissioning costs (spent fuel storage, site restoration, etc.) may be included in the cost estimate if desired; however, the cost of decommissioning, as defined by 10 CFR 50.2, should be listed separately. As a separate item, the cost of placing and maintaining the facility in safe storage should be identified, along with a plan to ensure that sufficient funds will be available for this purpose, if necessary, until such time as the radioactively contaminated material is placed in an authorized waste disposal site. The reviewer should note that, as with the PSDAR, 10 CFR 50.82(a)(8)(iii) requires a licensee to provide a SSCE within 2 years following permanent cessation of operations. If the cost estimate provided with the PSDAR was an SSCE, then this requirement has been satisfied.

Licensees who plan to use a period of storage or surveillance (SAFSTOR) are required by 10 CFR 50.82(a)(8)(iv) to provide a means of adjusting cost estimates and associated funding levels over the period of storage or surveillance. If a licensee intends to use a period of SAFSTOR, the reviewer should ensure that the licensee has included a description of its means of adjustment with its estimate of expected costs. The reviewer should determine whether the means described by the licensee provides adequate assurance that funds will be available for decommissioning activities at the time they are needed.

Table 5. Decommissioning Cost Distribution by Time Period—Reference PWR [a]

| Decommissioning Option | Decommissioning Cost (2000 $millions) [b] | | | | |
	Period 1 Planning & Preparation	Period 2 Plant Deactivation	Period 3 Safe Storage Operations	Period 4 Dismantle- ment	Total
DECON					
Period Years	2.5	0.6	6.3	1.7	11.1
Period Cost	14.3	56.9	10.8	151.7	233.7
SAFSTOR					
Period Years	2.5	0.6	57.7	1.7	62.5
Period Cost	14.3	56.9	144.3	148.5	364.0

[a] NUREG/CR-5884 (Ref. 5)
[b] Includes an assumed 25% contingency cost. SAFSTOR2 decommissioning option is assumed.

A.   **Cost Estimate Using Minimum Financial Assurance Funding Amount Method**

(1)   **Review Responsibilities**

Primary—Cognizant Project Manager, Project Directorate responsible for the reactor, Division of Licensing Project Management, Office of Nuclear Reactor Regulation, as assigned.

Secondary—Financial Reviewer, Financial and Regulatory Analysis Section, Reactor Policy and Rulemaking Branch, Division of Regulatory Improvement Programs, Office of Nuclear Reactor Regulation, or as assigned.

15

Table 6. Decommissioning Cost Distribution by Time Period -- Reference BWR [a]

Decommissioning Option	Decommissioning Cost (2000 $ millions) [b]				
	Period 1 Planning & Preparation	Period 2 Plant Deactivation	Period 3 Safe Storage Operations	Period 4 Dismantle-ment	Total
DECON					
Period Years	2.5	1.2	3.4	1.7	8.8
Period Cost	14.8	76.1	7.2	243.2	341.3
SAFSTOR					
Period Years	2.5	1.2	57.1	1.7	62.5
Period Cost	14.8	76.1	189.2	242.0	522.1

[a] NUREG/CR-6174 (Ref. 6)
[b] Includes an assumed 25% contingency cost. SAFSTOR2 decommissioning option is assumed.

(2) Areas of Review

This SRP directs the staff's review of the cost estimate that is required by
10 CFR 50.82(a)(4)(i) to be included in the PSDAR submitted prior to or within 2 years following permanent cessation of operations. The intent of this estimate of expected costs is to provide the NRC with an up-to-date cost estimate using the minium financial assurance funding amount method (10 CFR 50.75(c), the same method the licensee used in the submittal for establishing a fund for decommissioning as required by 10 CFR 50.75(b). This estimate will have been revised periodically during operation and may have been used in preparing the preliminary cost estimate.

(3) Acceptance Criteria

The acceptance criteria are based on regulations set out in 10 CFR 50.82(a)(4)(i). The regulations require that, within 2 years following permanent cessation of operations, the licensee shall submit a PSDAR to the NRC with a copy to the affected State or States. The report must include, among other things, an estimate of expected costs.

The acceptance criterion for the cost estimate is that the estimate at least equals the minimum financial assurance funding amount defined in 10 CFR 50.75(c) unless otherwise adequately justified. Only those costs contained in the description of decommissioning, as defined in 10 CFR 50.2, may be used to determine if the estimate at least equals the minimum funding requirement of 10 CFR 50.75(c). Therefore, the estimate should separate costs into categories that enable the reviewer to identify whether or not each listed item fits within the definition of decommissioning costs.

(4) Review Procedures

The reviewer will use the following process to determine that the submitted estimate of expected costs considers, in adequate detail, all major factors that could affect the cost to decommission.

The reviewer should verify that the procedure for calculating the MFA funding amount has been followed in determining the estimate of expected costs (see Section B.1). The reviewer should confirm that the cost estimate is provided in current year (estimate year) dollars, using disposal cost adjustment factors from the most recent revision of NUREG-1307, and that the factors affecting the funding algorithm calculation are verifiable.

The reviewer should confirm that the following information is provided and that all items are reasonable:

- Reactor thermal power rating

- Reactor type (PWR/BWR)

- Cost escalation factors (including an acceptable method of inflation adjustment; Section B.1 provides an acceptable method of allowing for escalation of costs due to inflation in unit costs of labor, energy (transportation), and waste burial).

(5) Evaluation Findings

Using the acceptance criteria in C.2.A(3) and the review procedure in C.2.A(4) of this section as a basis, the NRC reviewer shall verify that sufficient information has been provided to satisfy the requirements of the (10 CFR 50.82(a)(4)(i)). The estimate of expected costs shall be considered deficient if the decommissioning cost estimate is less than the financial assurance amount required by 10 CFR 50.75(c) and adequate justification is not provided. If deficiencies are discovered, the reviewer should provide this information to the NRC project manager for the plant. The NRC project manager will inform the licensee in writing of the deficiencies that must be corrected before major decommissioning activities can begin. The reviewer documents the findings of his/her review of the estimate of expected costs in a memorandum. The memorandum should be forwarded for inclusion in the review of the licensee's PSDAR.

(6) Implementation

The method described in this SRP will be used by the staff in evaluating conformance with the Commission's regulations, except when the licensee proposes an acceptable alternative for complying with specified portions of the regulations.

Table 7. Estimate of Expected Costs—PWR DECON [a]

Decommissioning Activity	Decommissioning Cost (2000 $millions) [b]				
	Period 1 (2.5 Years)	Period 2 (0.6 Years)	Period 3 (6.3 Years)	Period 4 (1.7 Years)	Duration (11.1 Years)
	Planning & Preparation	Plant Deactivation	Safe Storage Operations	Dismantle-ment	Total Cost
Radioactive Component Removal	0.0	0.7	0.0	11.8	12.5
Decontamination and Dismantlement	0.0	22.5	0.0	10.4	32.9
Management and Support	14.3	14.7	10.8	40.5	80.2
LLW Packaging	0.0	0.2	0.0	3.5	3.6
LLW Shipping	0.0	1.5	0.0	4.3	5.8
LLW Burial/Waste Vendor	0.0	17.3	0.0	81.3	98.5
Total Cost	14.3	56.9	10.8	151.7	233.6

[a] NUREG/CR-5884
[b] Assumes a 25% contingency cost.

Table 8. Estimate of Expected Costs—BWR DECON [a]

Decommissioning Activity	Decommissioning Cost (2000 $millions) [b]				
	Period 1 (2.5 Years)	Period 2 (1.2 Years)	Period 3 (3.4 Years)	Period 4 (1.7 Years)	Duration (8.8 Years)
	Planning & Preparation	Plant Deactivation	Safe Storage Operations	Dismantle-ment	Total Cost
Radioactive Component Removal	0.0	1.2	0.0	6.6	7.8
Decontamination and Dismantlement	0.0	20.8	0.0	15.8	36.6
Management and Support	14.8	34.7	7.2	40.0	96.8
LLW Packaging	0.0	0.2	0.0	5.5	5.7
LLW Shipping	0.0	1.1	0.0	0.4	1.5
LLW Burial/Waste Vendor	0.0	18.1	0.0	174.8	192.8
Total Cost	14.8	76.1	7.2	243.2	341.3

[a] NUREG/CR-6174
[b] Assumes a 25% contingency cost.

Table 9. Estimate of Expected Costs—PWR SAFSTOR [a]

Decommissioning Activity	Decommissioning Cost (2000 $millions) [b]				
	Period 1 (2.5 Years)	Period 2 (0.6 Years)	Period 3 (57.7 Years)	Period 4 (1.7 Years)	Duration (62.5 Years)
	Planning & Preparation	Plant Deactivation	Safe Storage Operations	Dismantlement	Total Cost
Radioactive Component Removal	0.0	0.7	0.0	11.8	12.5
Decontamination and Dismantlement	0.0	22.5	1.2	9.2	32.9
Management and Support	14.3	14.7	142.5	40.4	212.0
LLW Packaging	0.0	0.2	0.1	3.4	3.6
LLW Shipping	0.0	1.5	0.0	4.3	5.8
LLW Burial/Waste Vendor	0.0	17.3	0.4	79.4	97.0
Total Cost	14.3	56.9	144.3	148.5	363.9

[a] NUREG/CR-5884
[b] Assumes a 25% contingency cost. SAFSTOR2 decommissioning option is assumed.

Table 10. Estimate of Expected Costs—BWR SAFSTOR [a]

Decommissioning Activity	Decommissioning Cost (2000 $millions) [b]				
	Period 1 (2.5 Years)	Period 2 (1.2 Years)	Period 3 (57.1 Years)	Period 4 (1.7 Years)	Duration (62.5 Years)
	Planning & Preparation	Plant Deactivation	Safe Storage Operations	Dismantlement	Total Cost
Radioactive Component Removal	0.0	1.2	0.0	6.6	7.8
Decontamination and Dismantlement	0.0	20.8	0.7	15.1	36.6
Management and Support	14.8	34.7	188.2	41.6	279.3
LLW Packaging	0.0	0.2	0.0	5.5	5.7
LLW Shipping	0.0	1.1	0.0	0.4	1.5
LLW Burial/Waste Vendor	0.0	18.1	0.3	172.8	191.1
Total Cost	14.8	76.1	189.2	242.0	522.1

[a] NUREG/CR-6174
[b] Assumes a 25% contingency cost. SAFSTOR2 decommissioning option is assumed.

B.   Site-Specific Cost Estimate

The estimate of expected decommissioning costs required for the PSDAR can be the same as the site-specific cost estimate required by 10 CFR 50.82(a)(8)(iii). The site-specific cost estimate is a detailed assessment that incorporates the cost impact of site-specific factors. The site-specific estimate is discussed in Regulatory Position 3.

A site-specific cost estimate is required by 10 CFR 50.82(a)(8)(iii) to be submitted within 2 years following permanent cessation of operations. This cost estimate may be included with the PSDAR (10 CFR 50.82(a)(4)(i)). In addition, a licensee may submit a certification amount of funds for decommissioning based on a site-specific cost estimate that is equal to or greater than the amount calculated in the formula in 10 CFR 50.75(c)(1) or (2) when a higher funding level is desired. If the amount of the site-specific cost estimate is less than the certification formula amount, a licensee must provide adequate justification for the difference.

The SSCE is a very detailed assessment that incorporates the cost impact of site-specific factors. Because the SSCE that may be submitted with the PSDAR can be used to satisfy the requirement for a SSCE in 10 CFR 50.82(a)(8)(iii), the same review process should be used. The reviewer is referred to the Acceptance Criteria and Review Procedures that are provided in Section 3.

3.   SITE-SPECIFIC COST ESTIMATE

A SSCE is required by 10 CFR 50.82(a)(8)(iii) within 2 years following permanent cessation of operations. It may be included with the PSDAR (10 CFR 50.82(a)(4)(i)). The SSCE is intended to be based on a detailed analysis of the decommissioning costs required to safely dismantle and decontaminate the facility and site to meet the criteria for license termination. The SSCE submitted to the NRC may summarize the results of the detailed analyses with the underlying detail submitted as supplementary information. The summary data should be sufficiently detailed to demonstrate that the licensee has considered all significant decommissioning costs, and should reference the detailed cost estimate.

Licensees who plan to use a period of storage or surveillance (SAFSTOR) are required by 10 CFR 50.82(a)(8)(iv) to provide a means of adjusting cost estimates and associated funding levels over the period of storage or surveillance. If the time period covered by the updated SSCE includes a period of SAFSTOR, the reviewer should ensure that the licensee has included a description of its means of adjusting its SSCE. The reviewer should determine if the means described by the licensee provides adequate assurance that funds will be available for decommissioning activities at the time needed.

(1)   Review Responsibilities

Primary—Cognizant Project Manager, Project Directorate, Division of Licensing Project Management, Office of Nuclear Reactor Regulation, or Office of Nuclear Materials Safety and Safeguards depending on when submitted.

Secondary—Financial Reviewer, Financial and Regulatory Analysis Section, Reactor Policy and Rulemaking Branch, Division of Regulatory Improvement Programs, Office of Nuclear Reactor Regulation, or Office of Nuclear Material Safety and Safeguards.

(2)   Areas of Review

This SRP directs the staff's review of the SSCE that is required by 10 CFR 50.82(a)(8)(iii) within 2 years following permanent cessation of operations. The intent of this cost estimate is to provide the NRC with a detailed assessment that incorporates the cost impact of site-specific factors. Additionally, site-specific estimates may be submitted pursuant to 10 CFR 50.75(b) provided they are equal to or greater than the amount required by 10 CFR 50.75(c).

(3)   Acceptance Criteria

The acceptance criteria are based on regulations set out in 10 CFR 50.82(a)(8)(iii). The regulations require that within 2 years following permanent cessation of operations, if the licensee has not already submitted a SSCE with the PSDAR (10 CFR 50.82(a)(4)(i)).

To ensure that the cost estimate is site-specific and that all significant decommissioning costs have been considered, a SSCE should include the following items:

- A description of the decommissioning cost estimating methodology
- A description of the overall decommissioning project
- A summary decommissioning cost estimate by major activity and phase
- A schedule of the major decommissioning activities
- A summary of the radiological D&D management with support staff levels
- An estimate of the radioactive waste volume

(4)   Review Procedures

The reviewer will use the following process to determine that the submitted SSCE considers, in adequate detail, all major site-specific factors that could affect the cost to decommission, and to ensure that the SSCE appears reasonable.

The reviewer should compare the SSCE with the minimum decommissioning financial assurance requirement amount derived per the algorithm discussed in Section B.1 (10 CFR 50.75(c)). If the SSCE is less than the amount derived from the algorithm in 10 CFR 50.75(c) and adequate justification is not provided, the reviewer should provide this information to the NRC project manager for the plant. As discussed, the NRC project manager will inform the licensee in writing of additional information needed to resolve the deficiency.

The reviewer should first review the SSCE to determine if it is sufficiently detailed to allow the reviewer to make an assessment of its adequacy. If the reviewer is unable to find each of the detailed items, then the reviewer will need to make a determination as to whether enough information has been provided to evaluate each of the six items discussed under

Acceptance Criteria (above). If there is not sufficient information, the NRC reviewer will inform project manager, who will inform the licensee in writing of the additional information needed to resolve the deficiency.

1. The reviewer should confirm that the following information is provided:

a.   A description of the decommissioning cost estimating methodology

The reviewer should check for the following items to ensure that the licensee's description of the decommissioning cost methodology is complete.

- The decommissioning option/method—The reviewer should identify the decommissioning option/method that the licensee is planning to use. The decommissioning options generally available are DECON, SAFSTOR, or some combination thereof. Section A of this SRP describes each of these options. If the chosen option/method will result in completion of decommissioning more than 60 years after cessation of operations, identification and assessment of the factors causing this delay should be presented. Acceptable factors from 10 CFR 50.82(a)(3) include unavailability of waste disposal capacity and site-specific factors, such as the presence of other nuclear facilities at the site.

- A discussion of the methodology used to derive the cost estimates—The reviewer should identify the methodology used to develop the generic cost estimate. The most common methodology used to develop decommissioning cost estimates is the unit cost factor approach, which is the methodology utilized in the NRC reports mentioned above and the methodology developed by the Atomic Industrial Forum (now the Nuclear Energy Institute) for use by nuclear power plant licensees (AIF/NESP-036). Other methodologies, such as activity-based cost estimates, are acceptable.

b.   A description of the overall decommissioning project

The reviewer should check to ensure that the licensee has provided a detailed work breakdown for all the activities to be performed, including planning and preparation. The reviewer should specifically check that the following activities have been included:

- Planning and preparation
- Characterization survey of facility and site
- Disposal of ionexchanger resins
- Removal, radiological decontamination, and packaging of spent fuel racks
- Concentration and shipment of boron waste
- Radiological decontamination of systems using chemical cleaning methods
- Draining and processing of spent fuel pool water
- Removal of spent fuel pool cooling system
- Removal and packaging of reactor pressure vessel (RPV) internals

- Radiological decontamination and closure of RPV
- Removal of contaminated cranes
- Radiological decontamination, removal, and packaging of spent fuel pool liner
- Removal of reactor coolant system (RCS) piping and equipment
- Removal of pressurizer
- Removal of steam generators
- Removal of control rod drive system
- Segmentation and packaging of reactor pressure vessel
- Removal of bioshield shield
- Removal of turbine generator(s)
- Removal of turbine condenser(s)
- Removal of moisture separator reheaters
- Removal of feedwater heaters
- Removal of feedwater condensate system
- Removal of feedwater pumps/turbine drives
- Radiological decontamination and removal of floor drains
- Vacuuming or washing or other radiological decontamination of surfaces
- Removal of contaminated concrete
- Removal of heating, ventilation, and air conditioning ducts and equipment
- Removal of other contaminated systems (list each system)
- Remediation/removal of surface and groundwater
- Remediation/removal of contaminated soils
- Final survey
- LLW packaging, shipping, and burial charges, including LLW processing fees by waste vendors
- Shipment and processing or storage of greater-than-Class C waste

If the decommissioning project includes SAFSTOR periods (longer than about 5 years), the reviewer should also check that the schedule includes the following activities and labor requirements were included:

- Removal of any LLW that is ready to be shipped
- Deenergizing or deactivating specific systems
- Reconfiguration of ventilation systems and fire protection systems for use during the storage period
- Maintenance of any systems critical to final dismantlement during the storage period

23

- Mobilization of additional personnel at the end of the SAFSTOR period to begin the active phase of decommissioning work

The reviewer should also check for the following information:

- A summary of the inventory of contaminated systems and components requiring radiological decontamination and/or decommissioning (Table 11 provides an example of a contaminated equipment and piping inventory for the reference PWR and reference BWR (see NUREG/CR-5584 and NUREG/CR-6174)). The reviewer should compare the inventory provided with Table 11 in order to make a judgment regarding the reasonableness of the inventory.

Table 11. Example of Inventory for Contaminated Equipment and Piping

Equipment Category[a]	Reference PWR Length of Piping in Feet or Number of Items in Each Category	Reference BWR Length of Piping in Feet or Number of Items in Each Category
Piping diameter > 3 inches	15,110	55,654
Piping diameter ≤ 3 inches	34,631	66,160
Valves > 3 inches	235	1,103
Valves ≤ 3 inches	779	7,962
Tanks of all sizes	76	80
Pumps > 100 pounds	43	87
Pumps ≤ 100 pounds	2	8
Heat exchangers > 100 pounds	25	16
Heat exchangers ≤ 100 pounds	0	0
Electrical components > 100 pounds	69	0
Electrical components ≤ 100 pounds	34	0
Miscellaneous components > 100 pounds	13	1,323
Miscellaneous components ≤ 100 pounds	26	282
Large piping hanger, for pipes > 4 inches in diameter	2,204	5,000
Small piping hanger, for pipes ≤ 4 inches in diameter	10,608	7,500

[a] The equipment categories shown here are arbitrary. Any reasonable method of categorization is acceptable.

- An identification of the rooms and/or areas in the facility that need to be decontaminated (this information may have been either submitted by the licensee either as maps or provided in tables). Table 12 provides a table example of an inventory of concrete and metal surfaces requiring radiological decontamination/removal for the reference PWR and reference BWR. The reviewer should compare the inventory provided with Table 12 in order to make a judgment regarding the reasonableness of the inventory.

Table 12. Example of Inventory for Concrete and Metal Surfaces Requiring Decontamination/Removal

Reference PWR (DECON and SAFSTOR)				
Building or Location	Area of Concrete Decontaminated (ft^2)	Volume of Concrete Removed (ft^3)	Area of Metal Surfaces Decontaminated (ft^2)	Volume of Metal Surfaces Removed (ft^3)
Fuel Building	22,864	548	15,428	161
Containment Building	127,124	433	4,690	49
Auxiliary Building	43,860	819	0	0
Reference BWR (DECON and SAFSTOR)				
Building or Location	Area of Concrete Decontaminated (ft^2)	Volume of Concrete Removed (ft^3)	Area of Metal Surfaces Decontaminated (ft^2)	Volume of Metal Surfaces Removed (ft^3)
Reactor	30,537	1,304	33,906	541
Rad Waste/Control Building	21,711	388	1,526	16
Turbine Generator Building	8,042	123	1,526	16

- A summary description, based on the decommissioning records required by 10 CFR 50.75(g), of events occurring during operation involving the spread of contamination in and around the facility, equipment, or site, such that significant contamination remained after any cleanup procedures were carried out. Records of events that may have spread contamination into inaccessible areas or resulted in possible seepage into porous materials must be maintained. The decommissioning records must include as-built drawings and modifications to structures and equipment in restricted areas where radioactive materials were used or stored, and the locations of areas of possible inaccessible contamination, such as buried pipes. These records are intended to provide a historical record of the location, use, and spread of radioactive materials that can be used to guide decommissioning efforts.

Although the requirements described in 10 CFR 50.75(g) for keeping records of spills or other unusual occurrences are outside the scope of this SRP, the reviewer should ensure that the licensee has evaluated the anticipated extent of contamination on the facility and site based on information available in the decommissioning files. This description need not be a detailed discussion but should describe known instances of releases of contaminated materials into the facility and the external environment, as well as the possible impact on decommissioning. The licensee's discussion should include an evaluation of the historical use and location of radioactive materials at the site with an assessment of their impact on decommissioning costs.

The record-keeping requirements of 10 CFR 50.75(g) became effective on July 27, 1988. As a result, events that occurred before the effective date may not be included in the licensee's decommissioning records. Therefore, for plants with operating histories prior to July 1988, the reviewer should determine whether the licensee evaluated the plant's operating history and the modifications made to its facility, equipment, and site to assess their impact on decommissioning costs.

- A summary of available characterization information on known and/or suspected environmental contamination (soil, groundwater, and surface water). The reviewer should look for the identification of known environmental contamination (including soil, groundwater, surface water, etc.). The files that are required by 10 CFR 50.75(g) include records of spills or other unusual occurrences involving the spread of contamination in and around the facility, equipment, or site; as-built drawings and modifications of structures and equipment in restricted areas where radioactive materials are used and/or stored; locations of possible inaccessible contamination such as buried pipes.

- Records of the cost estimates performed for the decommissioning funding plan or the amount certified for decommissioning; and records of the funding method used for assuring funds if either a funding plan or certification is used.

- A summary description of structures or equipment in the restricted area where radioactive materials were used or stored, as well as the locations of possible inaccessible contamination.

c.   A summary decommissioning cost estimate by major activity and phase

- The reviewer should confirm that the cost estimate accounts for the entire decommissioning work scope, but not for items that are outside the scope of the decommissioning process such as the maintenance and storage of spent fuel in the spent fuel pool, the design or construction of spent fuel dry storage facilities, or other activities not directly related to the long-term storage, radiological D&D of the facility, or radiological decontamination of the site. If non-decommissioning cost items are included in the SSCE, these items should be identified separately. The SSCE should provide costs for each of the following, or similar, major activities and phases:

    - Major radioactive component removal—reactor vessel and internals, steam generators, pressurizers, large bore reactor coolant system piping, and other large components that are radioactive to a comparable degree, as defined in 10 CFR 50.2

    - Radiological D&D—removal of remaining radioactive plant systems, including radiological decontamination

    - Management and support—labor costs of support staff and decommissioning contractor's staff, energy costs, regulatory costs, small tools, insurance, etc.

    - LLW packaging—placing LLW in packages

    - LLW shipping—shipping LLW to waste vendors/burial site

- LLW burial/waste vendor—LLW burial charges, including LLW processing fees by waste vendors

- Contingency

If the SAFSTOR option is being used, the cost categories should also be segregated into the following:

- Pre-decommissioning engineering and planning/plant deactivation— all activities from engineering and planning through defueling and layup to completing the placement of the reactor into permanent shutdown condition

- Extended safe storage operations—safe storage monitoring of the facility until dismantlement begins (if storage or monitoring of spent fuel is included in the cost estimate, it should be shown separately)

- Final Radiological D&D—radiological D&D of radioactive systems and structures required for license termination, including demolition for the purposes of reducing residual radioactivity if demolition of uncontaminated structures and site restoration activities are included in the cost estimate, they should be shown separately

Tables 7 through 10 provide decommissioning cost estimates by decommissioning activities listed in Section 3(4)1c and time periods for the NRC reference PWR and reference BWR (see NUREG/CR-5884 and NUREG/CR-6174), respectively. The reviewer should compare the SSCE with the cost values provided in Tables 7 through 10 to make a judgment of the reasonableness of the SSCE, recognizing the difference between the reactor for which the SSCE was developed and the reference reactors.

- An estimate of the cost necessary to place and maintain the reactor in a safe storage condition if such action becomes necessary

- A description of how the contingency costs are calculated

- A description of how inflation is accounted for in the cost estimate—The reviewer should confirm that the cost estimate is provided in current year (estimate year) dollars and that escalation of the LLW disposition costs is considered separately from the general inflation rate applicable to labor, material, and energy costs. The reviewer should be aware of escalation rates used in the current revision of NUREG-1307.

- A schedule showing the amount of decommissioning funds currently available, the accumulation of additional funds, and the expenditure of the decommissioning funds

- The assumptions, references, and bases for unit costs that were used in developing the estimates

d.   A schedule of decommissioning activities

The reviewer should check to ensure that the schedule includes a work breakdown decommissioning activities (as discussed previously), periods of interim safe storage, labor requirements (person-hours), and key milestones.

e.   Radiological D&D management— Support and DOC staffing levels

The reviewer should check to ensure that the licensee has estimated staffing levels, labor requirements, and labor costs for each decommissioning phase (including periods of SAFSTOR, if applicable). Radiological D&D staff requirements may vary from site to site, depending on management. For this reason, the reviewer should determine if labor rates were adjusted for escalation and region accordingly.

f.   Radioactive waste information

The reviewer should determine if the licensee submitted estimates of radioactive waste volumes that are expected to be generated during decommissioning, assuming no volume reduction. Radioactive waste (radwaste) volumes should be identified by waste class. In addition, the reviewer should identify if the licensee submitted plans for radwaste disposition, including radwaste disposal sites to be used, if available. If the licensee has specified that a vendor will process the waste, then the radwaste information after processing should be available to show the results of the waste minimization. The licensee may also have included descriptions of the methods and technologies employed to achieve the improved waste characteristics.

2.   The reviewer should assess the reasonableness of submitted SSCEs and compare the information that was submitted with the information that is provided in this section for the reference PWR and BWR using the following process.

a)   The reviewer should compare the information presented in this section for the referenced PWR or BWR with the level of detail provided in the SSCE. The reviewer should check to see if there are items that appear to be significantly less than the amounts given in the following tables (taking into account the differences in plant sizes or decommissioning techniques) or that are significantly out of proportion. If the numbers are significantly different or out of proportion, before determining that the SSCE is deficient, the reviewer should check for an explanation or reason that might account for the difference.

b)   The reviewer should compare the cost estimates with detailed analyses as the reevaluated analyses of decommissioning of the NRC reference PWR and the reference BWR (see NUREG/CR-5884 and NUREG/CR-6174). Summaries of reports to be used for this comparison are presented below for a PWR undergoing the immediate dismantlement option (DECON) in Table 13 and for the safe storage option (SAFSTOR) in Table 14. Likewise for a BWR, Table 15 summarizes the DECON option and Table 16 summarizes the SAFSTOR option.

Table 13. Reference PWR Decommissioning Cost Distribution by Time Period— DECON

Decommissioning Activity	Decommissioning Cost (2000 $ thousands)				
	Period 1 (2.5 Years) Planning & Preparation	Period 2 (0.6 Years) Plant Deactivation	Period 3 (6.3 Years) Safe Storage Operations	Period 4 (1.7 Years) Dismantlement	Duration (11.1 Years) Total Cost
Radioactive Component Removal					
Removal of RPV Internals	0	743	0	0	743
Removal of Reactor Pressure Vessel	0	0	0	254	254
Steam Generator Direct Removal Costs	0	0	0	9,789	9,789
Steam Generator Cascading Costs	0	0	0	223	223
RCS Piping	0	0	0	35	35
Large Miscellaneous RCS Piping	0	0	0	36	36
Small Miscellaneous RCS Piping	0	0	0	67	67
RCS Insulation	0	0	0	0	0
Pressurizer	0	0	0	13	13
Pressurizer Relief Tank	0	0	0	9	9
Primary Pumps	0	0	0	51	51
Spent Fuel Racks	0	0	0	1,038	1,038
Biological Shield	0	0	0	272	272
Subtotal	0	743	0	11,787	12,530
Decontamination and Dismantlement					
Decontamination of Site Buildings	0	22,487	0	2,002	24,490
Removal of Contaminated Plant Systems	0	0	0	8,418	8,418
Subtotal	0	22,487	0	10,420	32,908
Management and Support					
Support Staff	942	9,433	2,992	5,323	18,689
DOC Staff	7,579	0	1,516	18,737	27,832
Consultants/Other Staff	0	0	0	190	190
Termination Survey Costs	0	0	0	1,916	1,916
Regulatory Costs	561	582	35	1,608	2,787
Special Tools and Equipment	5,216	0	0	0	5,216
Environmental Monitoring Costs	0	47	48	130	225
Laundry Services	0	496	92	1,456	2,044
Small Tools and Minor Equipment	0	15	0	411	426
Nuclear Liability Insurance	0	2,695	5,934	3,199	11,827
Property Taxes	0	0	89	240	329
DOC Mobilization/Demobilization Costs	0	0	0	4,144	4,144
Steam Generator Undistributed Costs	0	0	0	328	328
Chemical Decon	0	414	0	0	414
Plant Power Usage	0	1,011	59	2,771	3,840
Subtotal	14,298	14,693	10,764	40,453	80,208
LLW Packaging	0	167	0	3,464	3,631
LLW Shipping	0	1,518	0	4,323	5,841
LLW Burial/Waste Vendor	0	17,251	0	81,264	98,515
Total	14,298	56,859	10,764	151,712	233,632

Table 14. Reference PWR Decommissioning Cost Distribution by Time Period— SAFSTOR

Decommissioning Activity	Decommissioning Cost (2000 $thousands)				
	Period 1 (2.5 Years) Planning & Preparation	Period 2 (0.6 Years) Plant Deactivation	Period 3 (57.7 Years) Safe Storage Operations	Period 4 (1.7 Years) Dismantle-ment	Duration (62.5 Years) Total Cost
Radioactive Component Removal					
Removal of RPV Internals	0	743	0	0	743
Removal of Reactor Pressure Vessel	0	0	0	254	254
Steam Generator Direct Removal Costs	0	0	0	9,789	9,789
Steam Generator Cascading Costs	0	0	0	223	223
RCS Piping	0	0	0	35	35
Large Miscellaneous RCS Piping	0	0	0	36	36
Small Miscellaneous RCS Piping	0	0	0	67	67
RCS Insulation	0	0	0	0	0
Pressurizer	0	0	0	13	13
Pressurizer Relief Tank	0	0	0	9	9
Primary Pumps	0	0	0	51	51
Spent Fuel Racks	0	0	0	1,038	1,038
Biological Shield	0	0	0	272	272
Subtotal	0	743	0	11,787	12,530
Decontamination and Dismantlement					
Decontamination of Site Buildings	0	22,487	1,184	818	24,490
Removal of Contaminated Plant Systems	0	0	0	8,418	8,418
Subtotal	0	22,487	1,184	9,236	32,908
Management and Support					
Support Staff	942	9,433	68,187	5,323	83,884
DOC Staff	7,579	0	3,032	18,737	29,348
Consultant/Other Staff	0	0	0	190	190
Termination Survey Costs	0	0	0	1,916	1,916
Regulatory Costs	561	582	2,443	1,608	5,194
Special Tools and Equipment	5,216	0	0	0	5,216
Environmental Monitoring Costs	0	47	3,968	130	4,145
Laundry Services	0	496	990	1,438	2,925
Maintenance Allowance	0	0	1,402	0	1,402
Small Tools and Minor Equipment	0	15	0	411	426
Nuclear Liability Insurance	0	2,695	54,329	3,199	60,223
Property Taxes	0	0	7,348	240	7,588
DOC Mobilization/Demobilization Costs	0	0	0	4,144	4,144
Steam Generator Undistributed Costs	0	0	0	328	328
Chemical Decon	0	414	0	0	414
Plant Power Usage	0	1,011	847	2,771	4,629
Subtotal	14,298	14,693	142,546	40,435	211,972
LLW Packaging	0	167	105	3,360	3,631
LLW Shipping	0	1,518	1	4,322	5,841
LLW Burial/Waste Vendor	0	17,251	422	79,355	97,028
Total	14,298	56,859	144,258	148,495	363,910

Table 15. Reference BWR Decommissioning Cost Distribution by Time Period—DECON

Decommissioning Activity	Decommissioning Cost (2000 $thousands)				
	Period 1 (2.5 Years) Planning & Preparation	Period 2 (1.1 Years) Plant Deactivation	Period 3 (3.4 Years) Safe Storage Operations	Period 4 (1.7 Years) Dismantle-ment	Duration (8.8 Years) Total Cost
Radioactive Component Removal					
RPV Internals	0	1,227	0	0	1,227
Reactor Pressure Vessel and Insulation	0	0	0	287	287
Sacrificial Shield	0	0	0	1,177	1,177
Recirculation Pumps	0	0	0	25	25
RCS Piping	0	0	0	1,635	1,635
RCS Piping Insulation	0	0	0	0	0
Main Turbine	0	0	0	382	382
Main Turbine Condenser	0	0	0	776	776
Moisture Separator Reheaters	0	0	0	188	188
Feedwater Heaters	0	0	0	104	104
Turbine Feed Pumps	0	0	0	21	21
Structural Beams, Plates, & Cable Trays	0	0	0	691	691
Spent Fuel Racks	0	0	0	1,298	1,298
Subtotal	0	1,227	0	6,585	7,812
Decontamination and Dismantlement					
Decontamination of Site Buildings	0	20,811	0	1,144	21,954
Removal of Contaminated Plant Systems	0	0	0	14,687	14,687
Subtotal	0	20,811	0	15,831	36,642
Management and Support					
Support Staff	1,336	26,154	2,253	7,689	37,432
DOC Staff	7,579	0	1,516	17,694	26,789
Consultantss/Other Staff	0	0	0	190	190
Termination Survey Costs	0	0	0	1,661	1,661
Regulatory Costs	561	677	136	959	2,333
Special Tools and Equipment	5,374	0	0	0	5,374
Environmental Monitoring Costs	0	92	26	130	247
Laundry Services	0	826	50	1,700	2,576
Small Tools and Minor Equipment	0	25	0	430	454
Nuclear Liability Insurance	0	5,016	3,202	3,199	11,417
DOC Mobilization/Demobilization Costs	0	0	0	4,144	4,144
Chemical Decontamination	0	328	0	0	328
Plant Power Usage	0	1,566	25	2,219	3,810
Subtotal	14,850	34,684	7,208	40,015	96,757
LLW Packaging	0	217	0	5,506	5,722
LLW Shipping	0	1,089	0	444	1,534
LLW Burial/Waste Vendor	0	18,064	0	174,781	192,845
Total	14,850	76,092	7,208	243,162	341,312

Table 16. Reference BWR Decommissioning Cost Distribution by Time Period—SAFSTOR

Decommissioning Activity	Decommissioning Cost (2000 $thousands)				
	Period 1 (2.5 Years) Planning & Preparation	Period 2 (1.2 Years) Plant Deactivation	Period 3 (57.1 Years) Safe Storage Operations	Period 4 (1.7 Years) Dismantle- ment	Duration (62.5 Years) Total Cost
Radioactive Component Removal					
RPV Internals	0	1,227	0	0	1,227
Reactor Pressure Vessel and Insulation	0	0	0	287	287
Sacrificial Shield	0	0	0	1,177	1,177
Recirculation Pumps	0	0	0	25	25
RCS Piping	0	0	0	1,635	1,635
RCS Piping Insulation	0	0	0	0	0
Main Turbine	0	0	0	382	382
Main Turbine Condenser	0	0	0	776	776
Moisture Separator Reheaters	0	0	0	188	188
Feedwater Heaters	0	0	0	104	104
Turbine Feed Pumps	0	0	0	21	21
Structural Beams, Plates, & Cable Trays	0	0	0	691	691
Spent Fuel Racks	0	0	0	1,298	1,298
Subtotal	0	1,227	0	6,585	7,812
Decontamination and Dismantlement					
Decontamination of Site Buildings	0	20,811	715	428	21,954
Removal of Contaminated Plant Systems	0	0	0	14,687	14,687
Subtotal	0	20,811	715	15,116	36,642
Management and Support					
Support Staff	1,336	26,154	101,702	9,171	138,364
DOC Staff	7,579	0	3,032	17,694	28,305
Consultants/Other Staff	0	0	0	190	190
Termination Survey Costs	0	0	0	1,661	1,661
Regulatory Costs	561	677	22,378	959	24,575
Special Tools and Equipment	5,374	0	0	0	5,374
Environmental Monitoring Costs	0	92	4,123	130	4,344
Laundry Services	0	826	981	1,843	3,651
Maintenance Allowance	0	0	1,465	0	1,465
Small Tools and Minor Equipment	0	25	0	430	454
Nuclear Liability Insurance	0	5,016	53,783	3,199	61,997
Property Taxes	0	0	0	0	0
DOC Mobilization/Demobilization Costs	0	0	0	4,144	4,144
Chemical Decontamination	0	328	0	0	328
Plant Power Usage	0	1,566	685	2,219	4,471
Subtotal	14,850	34,684	188,150	41,640	279,324
LLW Packaging	0	217	38	5,467	5,722
LLW Shipping	0	1,089	26	418	1,534
LLW Burial/Waste Vendor	0	18,064	270	172,768	191,103
Total	14,850	76,092	189,200	241,995	522,136

c)   The reviewer should compare the licensee's estimates with the tabulations of typical waste volumes, packaging costs, shipping costs, and burial costs for the reference PWR and the reference BWR (see NUREG/CR-5884 and NUREG/CR-6174) as shown in Tables 17 and 18 below. The most recent update of NUREG-1307 includes a discussion and analysis of recently-used waste volume reduction technologies. This analysis includes an option that assumes the utilization of waste vendors to process limited amounts of LLW that meets certain specifications. The updated NUREG-1307 also give the latest radioactive waste disposal unit costs and adjustment factors for waste burial at other licensed disposal sites.

Table 17. Typical Waste Burial Cost and Volumes—Reference PWR

Decommissioning Activity	Waste Volume (ft³)	Packaging Cost (2000 $ millions)	Shipping Cost (2000 $ millions)	Burial Cost (2000 $ millions)
DECON				
Removal of NSSS	123,700	1.38	5.22	49.00
Removal of Contaminated Plant	75,500	1.14	0.28	25.52
Decontamination of Site Buildings	72,500	1.00	0.28	19.25
Dry Active Waste	19,500	0.11	0.06	4.74
Total	291,200	3.63	5.84	98.52
SAFSTOR			-	
Removal of NSSS	123,700	1.38	5.22	47.71
Removal of Contaminated Plant	75,500	1.14	0.28	25.33
Decontamination of Site Buildings	72,500	1.00	0.28	19.25
Dry Active Waste	19,500	0.11	0.06	4.74
Total	291,200	3.63	5.84	97.03

Table 18. Typical Waste Burial Cost and Volumes—Reference BWR

Decommissioning Activity	Waste Volume (ft³)	Packaging Cost (2000 $ millions)	Shipping Cost (2000 $ millions)	Burial Cost (2000 $ millions)
DECON				
Removal of NSSS	293,200	3.04	1.37	113.85
Removal of Contaminated Plant	149,000	2.06	0.07	53.77
Decontamination of Site Buildings	57,700	0.42	0.08	16.48
Other Dry Active Waste	34,200	0.19	0.02	8.74
Total	534,100	5.72	1.53	192.84
SAFSTOR				
Removal of NSSS	293,200	3.04	1.37	113.81
Removal of Contaminated Plant	149,000	2.06	0.07	52.07
Decontamination of Site Buildings	57,700	0.42	0.08	16.48
Other Dry Active Waste	34,200	0.19	0.02	8.74
Total	534,100	5.72	1.53	191.10

d)   The reviewer should compare the licensee's schedule of decommissioning activities with the schedules shown in Figures 2 and 3 to ensure sufficient level of detail to determine the task scheduling, task durations, and labor requirements for decommissioning activities.

Figure 1. Schedule of Activities During Reference BWR Deactivation (Period 2)

Labor Hours	Qtr 1, 2000	Qtr 2, 2000	Qtr 3, 2000	Qtr 4, 2000	Qtr 1, 2001	Qtr 2, 2001	Qtr 3, 2001
58,41!	▮▬▬▬▬▬▬▬▬▬▬▬▬▬▬▬▬▬▬▬▬▮ Reactor Deactivation						
1,92(	▦ Conduct radiation survey for baseline for chemical decontamination of systems						
12,96(	▦▦▦▦ Perform chemical decontamination						
5,75(	▦▦▦ Deactivate support Systems						
21,64;	▦▦▦▦▦▦▦▦ Cut, remove, package internals						
2,01(	▦ Drain, decontaminate dryer separator pool						
4,03;	▦ Treat, release water from RPV, dryer separator pool						
10,08(	▦▦▦▦▦▦▦▦ Package radioactive wastes						

34

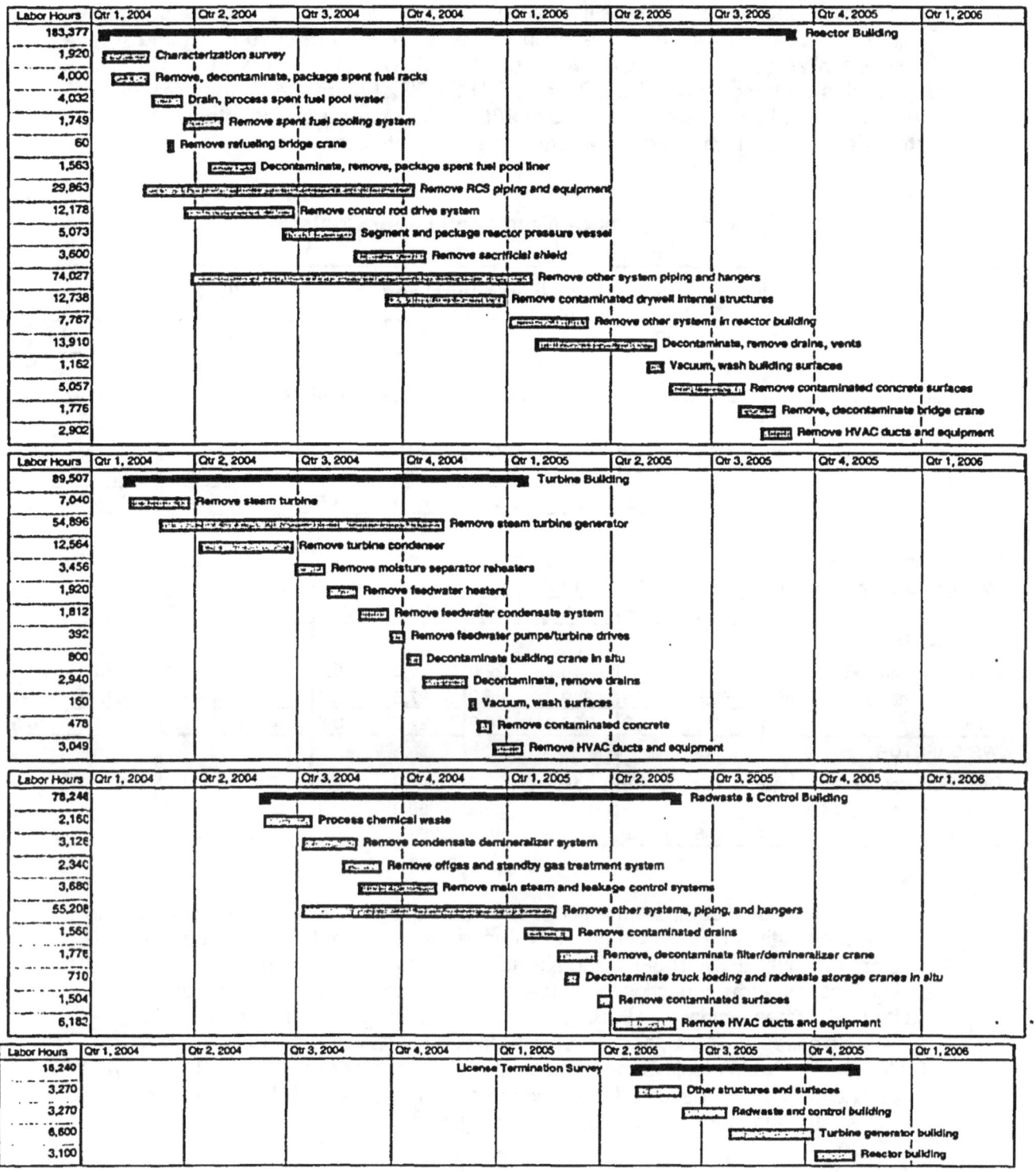

Figure 2. Schedule of Activities During Reference BWR Dismantlement (Period 4)

35

e) The reviewer should compare the licensee's estimated labor needs and labor costs by time period with those shown below in Table 19 for the reference PWR and reference BWR (see NUREG/CR-5884 and NUREG/CR-6174) for both decommissioning scenarios, DECON and SAFSTOR. Labor needs (in person-years per period) and labor costs (in millions of 2000 dollars) are grouped into two labor categories, decommissioning crews and management/support staff.

Table 19. Labor Needs and Labor Costs

| | Labor Needs (person-yrs) and Labor Costs (2000 $millions) | | | | | | | | | |
| | Period 1 | | Period 2 | | Period 3 | | Period 4 | | Total | |
	(Labor Need)	(Labor Cost)	(Labor Need)	(Labor Cost)	(Labor Need)	(Labor Cost)	(Labor Need)	(Labor Cost)	(Labor Need)	(Labor Cost)
PWR DECON										
Decommissioning Crews	0.0	0.0	16.0	23.2	0.0	0.0	122.0	22.2	138.0	45.4
Management/Support Staff	55.5	8.5	112.7	9.4	42.9	4.5	169.0	26.2	380.1	48.6
Total	55.5	8.5	128.7	32.7	42.9	4.5	290.9	48.4	518.1	94.1
PWR SAFSTOR										
Decommissioning Crews	0.0	0.0	16.0	23.2	2.1	1.2	119.9	21.0	138.0	45.4
Management/Support Staff	55.5	8.5	112.7	9.4	936.9	71.2	181.0	26.2	1,286.0	115.3
Total	55.5	8.5	128.7	32.7	938.9	72.4	300.9	47.2	1,424.0	160.8
BWR DECON										
Decommissioning Crews	0.0	0.0	16.7	22.0	0.0	0.0	168.7	22.4	185.4	44.5
Management/Support Staff	55.5	8.9	219.6	26.2	27.5	3.8	176.6	27.2	479.2	66.1
Total	55.5	8.9	236.3	48.2	27.5	3.8	345.3	49.7	664.6	110.5
BWR SAFSTOR										
Decommissioning Crews	0.0	0.0	16.7	22.0	1.3	0.7	167.3	21.7	185.4	44.5
Management/Support Staff	55.5	8.9	219.6	26.2	960.9	104.7	191.8	28.7	1,427.8	168.5
Total	55.5	8.9	236.3	48.2	962.2	105.4	359.2	50.4	1,613.2	213.0

f) The reviewer should compare the licensee's estimate of radwaste volumes with the approximate estimates made in the reevaluated analyses of the NRC reference reactors (see NUREG/CR-5884 and NUREG/CR-6174). Those analyses assumed no significant volume reductions and used waste containers, transportation and waste burial rates typical for 1993. The distribution range of waste burial volumes by waste classes A, B & C, and greater than class C (GTCC) are shown below in Table 20. The table displays the combined volume of classes B & C. All Class A and B & C wastes are assumed to be disposed at licensed LLW burial sites with GTCC waste being stored in a licensed geologic repository.

Table 20. Burial Volumes by Waste[a]

Waste Class	Reference PWR		Reference BWR	
	Volume (ft³)	Percent	Volume (ft³)	Percent
Class A	280,900	96.5	514,900	96.4
Class B&C	9,900	3.4	19,200	3.6
GTCC	400	0.13	200	0.04
Total	291,200	100.0	534,100	100.0

[a] Untreated (prior to volume reduction) volumes.

(5) Evaluation Findings

Using the acceptance criteria in C.3(3) and the review procedure in C.3(4) of this section as a basis, the NRC staff reviewer shall verify that sufficient information has been provided to satisfy the requirement of the underlying regulations (10 CFR 50.82(a)(8)(iii) or 10 CFR 50.75(b)). The SSCE shall be considered deficient if (1) the decommissioning cost estimate is less than the financial assurance amount required by 10 CFR 50.75(c) and adequate justification is not provided, (2) the reviewer cannot verify that all the information identified under the Acceptance Criteria has been provided, or (3) in the reviewer's judgment the SSCE submitted does not appear reasonable based on a comparison with the information provided from the reference PWR or BWR, considering the variation in plant sizes and decommissioning techniques. If deficiencies are discovered, the reviewer should provide this information to the NRC project manager for the plant. The NRC project manager will inform the licensee in writing of the additional information that is needed to ensure that the SSCE can be adequately evaluated. The reviewer documents the findings of his/her review of the SSCE in a memorandum to his/her branch chief with a copy to the NRC project manager for the plant.

(6) Implementation

The method described in this SRP will be used by the staff in evaluating conformance with the NRC's regulations, except when the licensee proposes an acceptable alternative for complying with specified portions of the regulations.

4. LICENSE TERMINATION PLAN UPDATED SITE-SPECIFIC COST ESTIMATE

According to 10 CFR 50.82(a)(9)(ii)(F), a licensee must submit "[a]n updated site-specific estimate of remaining decommissioning costs..." as part of an LTP. According to 10 CFR 50.82(a)(9)(i), among other things, the licensee must submit the LTP at least 2 years before termination of the license. The estimated remaining costs of decommissioning must be compared with the present funds set aside for decommissioning. The financial assurance instrument required per 10 CFR 50.75(b)(1) must be funded at least to the amount of the cost estimate. If there is a deficit in present funding, the LTP must indicate the means for ensuring adequate funds to complete the decommissioning. Information on the preparation of an LTP may be found in Regulatory Guide 1.179, "Standard Format and Content of License Termination Plans for Nuclear Power Reactors" and NUREG-1700, "Standard Review Plan for Evaluating Nuclear Power Reactor License Termination Plans." NUREG-1700, "Update of Site-Specific Costs" addresses the information necessary to support the cost estimate. The update of the site specific costs may be in summary form provided the supporting information had been previously submitted and is referenced. The supporting information may have been submitted as part of the SSCE or the expected cost estimated submitted with the PSDAR.

Licensees who plan to use a period of storage or surveillance (SAFSTOR) are required by 10 CFR 50.82(a)(8)(iv) to provide a means of adjusting cost estimates and associated funding levels over the period of storage or surveillance. If the time period covered by the updated SSCE includes a period of SAFSTOR, the reviewer should ensure that the licensee has included a description of its means of adjustment in the updated SSCE. The cost estimate reviewer should consult with a financial assurance reviewer to determine if the means described by the licensee provide adequate assurance that funds will be available for decommissioning activities at the time they are needed. Cost estimates associated with requests for license termination under restricted release conditions and for entombment will be handled on a case-by-case basis.

(1)   Review Responsibilities

Primary—Division of Waste Management, Office of Nuclear Material Safety and Safeguards

Secondary—Financial Reviewer, Financial and regulatory Analysis Section, Reactor Policy and Rulemaking Branch, Division of Regulatory Improvement Programs, Office of Nuclear Reactor Regulation, or as assigned.

(2)   Areas of Review

This SRP directs the staff's review of the "an updated site-specific estimate of remaining decommissioning costs" that is required by 10 CFR 50.82(a)(9)(ii)(F) as part of an LTP. The intent of this cost estimate is to provide the NRC with an up-to-date site-specific estimate of remaining decommissioning costs to terminate the license. A complete SSCE will have been submitted within 2 years following permanent cessation of operations.

(3)   Acceptance Criteria

In accordance with 10 CFR 50.82(a)(9)(i), a licensee must submit its LTP at least 2 years before termination of the license. The LTP submittal must be a supplement to the final safety analysis report (FSAR) or equivalent. In accordance with 10 CFR 50.82(a)(9)(ii)(F), the LTP must contain "an updated site-specific estimate of remaining decommissioning costs...."

The LTP cost estimate should contain, for those activities remaining to be completed, an updated, equally detailed version of the site-specific estimate previously submitted to and accepted by the NRC. The updated cost estimate in the LTP should include the following items:

- Estimated costs of remaining radiological decontamination activities
- Estimated costs of dismantling remaining contaminated equipment and structures

- Estimated costs for disposal of remaining radioactive waste

- Estimated final survey costs and license termination survey costs

- If the site is released for restricted use, the estimated costs for controls and a description of the financial assurance mechanisms used to ensure the availability of funds when they are needed

A licensee may include nondecommissioning costs in its LTP for information purposes. However, if the licensee does so, such costs should be clearly identified as costs in addition to decommissioning costs.

(4) Review Procedures

The reviewer will use the following process to determine that the submitted LTP decommissioning cost estimate considers, in adequate detail, all major factors that could affect the total remaining cost to decommission.

The reviewer should review the LTP decommissioning cost estimate to determine if it is sufficiently detailed to allow the reviewer to assess its adequacy. To make this assessment, the reviewer should confirm that the cost estimate is provided in current year (estimate year) dollars and that escalation of the LLW disposition costs is considered separately from the general inflation rate applicable to labor, material, and energy costs. The reviewer should be aware of the escalation rates used in the current revision of NUREG-1307. The reviewer should also confirm that the cost estimate accounts for the entire decommissioning work scope, but not for items that are outside the scope of the decommissioning process, such as the maintenance and storage of spent fuel in the spent fuel pool, the design or construction of spent fuel dry storage facilities, or other activities not directly related to the long-term storage, radiological D&D of the facility, or radiological decontamination of the site.

The reviewer should ensure that (1) the licensee has identified the remaining dismantlement activities that are necessary to complete the decommissioning of the facility/site, as required by 10 CFR 50.82(a)(9)(ii)(B), and (2) the licensee has identified site areas requiring remediation and has in place an organization to safely perform the remediation as required by 10 CFR 50.82(a)(9)(ii)(C). The licensee should have provided costs for each of the following cost elements identified below.

Cost Elements

- Cost assumptions used, including a contingency factor

- Major remaining decommissioning activities and tasks

- Estimated costs of radiological decontamination and removal of remaining radioactive equipment and structures

- Estimated costs of waste disposal, including applicable disposal site surcharges and transportation costs

- Estimated final survey costs

- Estimated total costs

The previous SRP for the SSCE gives further details on this analysis, including the specific information that should have been provided and descriptions of the type of information and anticipated values.

(5)   Evaluation Findings

Using the acceptance criteria in C.4(3) and the review procedures in C.4(4) of this section as a basis, the NRC staff reviewer shall verify that sufficient information has been provided to satisfy the requirements of the underlying regulations (10 CFR 50.82(a)(9)(ii)(F)). The LTP decommissioning cost estimate shall be considered deficient if any of the costs listed in the acceptance criteria are not adequately addressed. If deficiencies are discovered, the reviewer should provide this information to the NRC project manager for the plant. The NRC project manager will inform the licensee in writing of the additional information that is required by the regulations before major decommissioning activities can begin. The reviewer documents the findings of his/her review of the LTP decommissioning cost estimate in a memorandum to his/her branch chief with a copy to the NRC project manager for the plant. The review should be forwarded for inclusion in the LTP evaluation.

(6)   Implementation

The method described in this SRP will be used by the staff in evaluating conformance with the Commission's regulations, except when the licensee proposes an acceptable alternative for complying with specified portions of the regulations.

D. REFERENCES

AIF/NESP-036, "Guidelines for Producing Commercial Nuclear Power Plant Decommissioning Cost Estimates," Atomic Industrial Forum, Inc., May 1986.

Bureau of Labor Statistics, "*Monthly Labor Review*," Table 24, U.S. Department of Labor, Updated Periodically.

Bureau of Labor Statistics, "*Producer Price Index*," Table 6, U.S. Department of Labor, Updated Periodically.

DG-1085, "Standard Format and Content of Decommissioning Cost Estimates for Nuclear Power Reactors," Regulatory Guide, USNRC, November 2001.[1]

NUREG-0586, "Generic Environmental Impact Statement on Decommissioning of Nuclear Facilities, Supplement 1" USNRC, October 2002.[2]

NUREG-1307, "Report on Waste Disposal Charges: Changes in Decommissioning Waste Disposal Costs at Low-Level Waste Burial Facilities," Rev. 9, USNRC, September 2000.[2]

NUREG-1577, "Standard Review Plan on Power Reactor Licensee Financial Qualifications and Decommissioning Funding Assurance," Revision 1, USNRC, March 1999.[2]

NUREG-1700, "Standard Review Plan for Evaluating Nuclear Power Reactor License Termination Plans," NUREG-1700, USNRC, April 2000.[2]

NUREG/CR-0130, R. I. Smith, G. J. Konzek, and W. E. Kennedy, Jr.; "Technology, Safety and Costs of Decommissioning a Reference Pressurized Water Reactor Power Station" (Prepared for the U.S. NRC by Pacific Northwest Laboratory, Richland, Washington), June 1978 (Addendum 1, July 1979; Addendum 2, July 1983; Addendum 3, September 1984; Addendum 4, July 1988).[2]

NUREG/CR-0672, H. D. Oak et al., "Technology, Safety and Costs of Decommissioning a Reference Boiling Water Reactor Power Station" (prepared for the U.S. NRC by Pacific Northwest Laboratory, Richland, Washington), June 1980 (Addendum 1, July 1983; Addendum 2, September 1984; Addendum 3, July 1988; Addendum 4, December 1990).[2]

NUREG/CR-5884, G. J. Konzek et al., "Revised Analyses of Decommissioning for the Reference Pressurized Water Reactor Power Station," (prepared for the U.S. NRC by Pacific Northwest Laboratory, Richland, Washington), November 1995.[2]

---

[1] Single copies of regulatory guides may be obtained free of charge by writing the Reproduction and Distribution Services Section, OCIO, USNRC, Washington, DC 20555-0001, or by fax to (301) 415-2289, or by email to DISTRIBUTION@NRC.GOV. Copies of guides are available for inspection or copying for a fee from the NRC Public Document Room at 11555 Rockville Pike, Rockville, MD; the PDR's mailing address is USNRC PDR, Washington, DC 20555; telephone (301) 415-4737 or (800) 397-4209; fax (301) 415-3548; email PDR@NRC.GOV.

[2] Copies are available at current rates from the U.S. Government Printing Office, P.O. Box 37082, Washington, DC 20402-9328 (telephone (202) 512-1800); or from the National Technical Information Service by writing NTIS at 5285 Port Royal Road, Springfield, VA 22161; http://www.ntis.gov/ordernow; telephone (703)487-4650;. Copies are available for inspection or copying for a fee from the NRC Public Document Room at 11555 Rockville Pike, Rockville, MD; the PDR's mailing address is USNRC PDR, Washington, DC 20555; telephone (301) 415-4737 or (800) 397-4209; fax (301) 415-3548; email is PDR@NRC.GOV.

NUREG/CR-6174, R. I. Smith et al., "Revised Analyses of Decommissioning for the Reference Boiling Water Reactor Power Station," (Prepared for the U.S. NRC by Pacific Northwest National Laboratory, Richland, Washington), July 1996.[2]

Regulatory Guide 1.179, "Standard Format and Content of License Termination Plans for Nuclear Power Reactors," USNRC, January 1999.[1]

Regulatory Guide 1.185, "Standard Format and Content for Post-Shutdown Decommissioning Activities Report," USNRC, August 2000.[1]

NRC FORM 335 (9-2004) NRCMD 3.7	U.S. NUCLEAR REGULATORY COMMISSION	1. REPORT NUMBER (Assigned by NRC, Add Vol., Supp., Rev., and Addendum Numbers, if any.)
	BIBLIOGRAPHIC DATA SHEET *(See Instructions on the reverse)*	NUREG-1713

2. TITLE AND SUBTITLE	3. DATE REPORT PUBLISHED	
Standard Review Plan for Decommissioning Cost Estimates for Nuclear Power Reactors	MONTH	YEAR
	December	2004
	4. FIN OR GRANT NUMBER	

5. AUTHOR(S)	6. TYPE OF REPORT
Clayton L. Pittiglio	Final
	7. PERIOD COVERED *(Inclusive Dates)*

8. PERFORMING ORGANIZATION - NAME AND ADDRESS *(If NRC, provide Division, Office or Region, U.S. Nuclear Regulatory Commission, and mailing address; if contractor, provide name and mailing address.)*

Division of Regulatory Improvement Programs
Office of Nuclear Reactor Regulation
U.S. Nuclear Regulatory Commission
Washington, DC 20555-0001

9. SPONSORING ORGANIZATION - NAME AND ADDRESS *(If NRC, type "Same as above"; if contractor, provide NRC Division, Office or Region, U.S. Nuclear Regulatory Commission, and mailing address.)*

Same as 8 above.

10. SUPPLEMENTARY NOTES

11. ABSTRACT *(200 words or less)*

This Standard Review Plan (SRP) for decommissioning cost estimates provides guidance to Office of Nuclear Reactor Regulation (NRR) and Office of Nuclear Material Safety and Safeguards (NMSS) staff on how to evaluate each of the decommissioning cost estimates that are required to be provided by the power reactor licensees. The SRP includes guidance on evaluating decommissioning costs for both pressurized water reactors (PWRs) and boiling water reactors (BWRs). The SRP is divided into sections that are keyed to the sections in Regulatory Guide-1085, "Standard Format and Content of Decommissioning Cost Estimates for Nuclear Power Reactors," which was developed to provide guidance to licensees on decommissioning cost estimates. Each section of this NUREG is a separate SRP and presents the areas of review, acceptance criteria, review procedures, and evaluation findings for each of the decommissioning cost estimates required by 10CFR50.75 and 10CFR50.82.

12. KEY WORDS/DESCRIPTORS *(List words or phrases that will assist researchers in locating the report.)*	13. AVAILABILITY STATEMENT
Decommissioning, Cost Estimates, DECON, SAFSTOR, ENTOMB	unlimited
	14. SECURITY CLASSIFICATION
	(This Page) unclassified
	(This Report) unclassified
	15. NUMBER OF PAGES
	16. PRICE

NRC FORM 335 (9-2004)   PRINTED ON RECYCLED PAPER